Fokker Eindecker in action

By D. Edgar Brannon

Color by Don Greer

Illustrated by Joe Sewell & Randle Toepfer

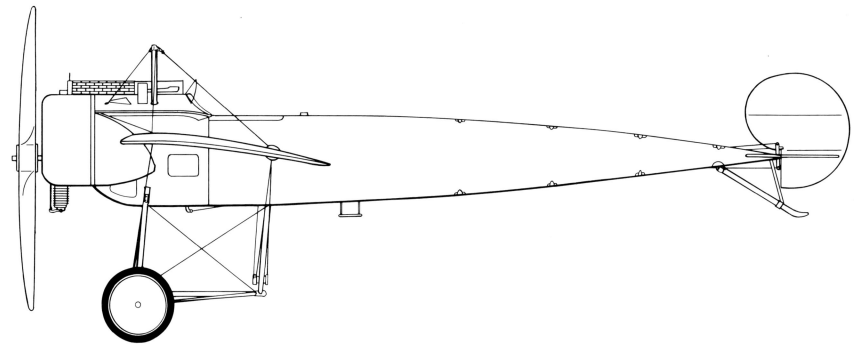

Aircraft Number 158

squadron/signal publications

Ernest Udet, flying a Fokker E.III (105/15), encounters a Royal flying Corps DH-2 over the Western Front during the Spring of 1916. The introduction of the DH-2 was one of the factors that led to the withdrawal of the Eindecker from service.

Acknowledgements

The Author wishes to thank the following: Frank H. Allen III for his support and assistance in screening photos of the Fokker E series at the San Diego Aerospace Museum; Peter M. Grosz for providing a number of photographs from his collection; Peter M. Bowers for his assistance in compiling the photos and captions used in this book; Lee Hilles for her editorial assistance; and my beautiful daughter Susan Brannon-Miles for preparation of the manuscript.

AUTHOR'S NOTE:

References. This action took place eighty years ago and revolves around Anthony Herman Gerard Fokker, a contractor to the German Army Air Service during the First World War. The ravages brought by Germany's defeat in both World Wars played havoc with the survival of any official German records relating to these years. A wealth of works have been written about the First World War. But, if you trim these down to those written about the war in the air, the sources are reduced considerably. Trim again to books written in or translated to English about German aerial activities and the number becomes very small. The Author has relied extensively on three references: Henri Hegener et al., *Fokker- The Man and the Aircraft*; Lamberton et al., *Fighter Aircraft of the 1914-1918 War*; and J.M. Bruce, *The Fokker Monoplanes*. Additional reference was drawn from: Robert Jackson, *Fighter Pilots of World War One*; and P.M. Grosz, *Fokker E-III*. The Author believes these references to be highly reliable. However, some conflicts in data exist and these are pointed out.

Dedication:

This work is dedicated to that group of volunteers, led by Frank H. Allen III, that are constructing a replica Fokker E.III at the San Diego Aerospace Museum.

COPYRIGHT 1996 SQUADRON/SIGNAL PUBLICATIONS, INC.
1115 CROWLEY DRIVE CARROLLTON, TEXAS 75011-5010
All rights reserved. No part of this publication may be reproduced, stored in a retrieval system or transmitted in any form by means electrical, mechanical or otherwise, without written permission of the publisher.

ISBN 0-89747-351-5

If you have any photographs of aircraft, armor, soldiers or ships of any nation, particularly wartime snapshots, why not share them with us and help make Squadron/Signal's books all the more interesting and complete in the future. Any photograph sent to us will be copied and the original returned. The donor will be fully credited for any photos used. Please send them to:

Squadron/Signal Publications, Inc.
1115 Crowley Drive
Carrollton, TX 75011-5010

Если у вас есть фотографии самолётов, вооружения, солдат или кораблей любой страны, особенно, снимки времён войны, поделитесь с нами и помогите сделать новые книги издательства Эскадрон/Сигнал ещё интереснее. Мы переснимем ваши фотографии и вернём оригиналы. Имена приславших снимки будут сопровождать все опубликованные фотографии. Пожалуйста, присылайте фотографии по адресу:

Squadron/Signal Publications, Inc.
1115 Crowley Drive
Carrollton, TX 75011-5010

軍用機、装甲車両、兵士、軍艦などの写真を所持しておられる方はいらっしゃいませんか？どの国のものでも結構です。作戦中に撮影されたものが特に良いのです。Squadron/Signal社の出版する刊行物において、このような写真は内容を一層充実し、興味深くすることができます。当方にお送り頂いた写真は、複写の後お返しいたします。出版物中に写真を使用した場合は、必ず提供者のお名前を明記させて頂きます。お写真は下記にご送付ください。

Squadron/Signal Publications, Inc.
1115 Crowley Drive
Carrollton, TX 75011-5010

Credits

Peter M. Bowers
Fokker Company, Holland
United States Air Force Museum.

Peter M. Grosz
San Diego Aerospace Museum

A collection of Fokker aircraft on the air park adjacent to the factory at Schwerin. The aircraft in the foreground carried the numbers LF210, LF211 and LF212. These are German Navy aircraft assigned to land based duty. The center aircraft is an E.IV, while those on either side are E.IIIs. (Fokker)

Introduction

Until shortly before the start of the First World War, Fokker's business efforts consisted of the construction of *Spin* (spider) aircraft, and the operation of a flying school for military and civil students. During this time, sales were sparse and the period 1910 to 1914 became known as Anthony Fokker's "years of adversity." During 1913, Fokker produced the first prototype of the E-Series monoplane, which he continued to produce through 1916.

This work covers Fokker's early years and his production of various versions of the Fokker *Eindecker* series, the E.I, E.II, E.III and E.IV. These aircraft introduced a new dimension to warfare -- combat in the air. At the time of its introduction, air-to-air combat was nearly nonexistent. After the Fokker monoplanes appeared, with their synchronized machine guns, air combat developed with a vengeance. This new era, that of "fighter aircraft," became at least as vicious as the fighting in the trenches and the average life expectancy of a pilot was six weeks.

Fokker was an extraordinary man: entrepreneur, inventor, innovator and crack pilot. He earned his pilot's certificate at the age of twenty, tested and certified in an airplane of his own design and construction. At the age of twenty-two he started his own company, teaching new pilots and manufacturing aircraft. His company still exists today as one of the leaders in aircraft design and manufacture.

Anthony Herman Gerard Fokker was born of Dutch parents in the Dutch East Indies (now known as Indonesia) on 6 April 1890. His father owned a large plantation, growing and exporting coffee. When Fokker was four, his father retired from business, sold his interests and relocated the family to Holland, settling in Haarlem and, being well-to-do, led a relaxed life.

Late in 1908, Anthony Fokker dropped out of **hoch schule** (high school -- equivalent to an undergraduate school in the U.S.) in the middle of his fourth year. His father later admitted that this was not all that bad, since he had done the same thing himself! The Dutch military called Fokker for compulsory service and this did not suit him at all! He went to Naarden, Holland for service but returned to Haarlem a short time later. A bad attitude and attempts to appear disabled to the military doctors caused him to be rejected from the military in short order.

Fokker quickly bounced back. He and a long-time friend, Fritz Cremer, invented a "puncture-proof" automobile tire. The pair was ready to go into immediate production; all they required was a sizable investment from their fathers. Herman Fokker hired a lawyer to look into the patentability of their invention and found that a French patent had been granted for just such a tire a short time earlier.

Fokker, then nineteen, "bummed" around Haarlem for the next several months. He worked

This Spin 3 was built and flown by Fokker in May of 1911. He received his flying certificate in this aircraft on 16 May 1911. Two years later, this same type of aircraft was purchased by the German military under the designation Militar 1 (M.1). (Grosz)

A Fokker Spin 3 at the Fokker flying school with the instructor and student ready for flight. (Bowers)

for Jan De Boer, learning to operate various metal-working machines, including lathes.

Fokker's father, concerned about his son's future, began looking into opportunities that would provide a business future for his son; a vocation where his mechanical skills could be applied. A brochure came to his attention that described an automotive engineering school at Bingen, Germany. Attending this school away from home would put Fokker on his own feet, or so his father thought, and provide a business future. Hindsight indicates that this decision was certainly a correct one.

Some amount of persuasion must have been necessary to talk young Fokker into leaving home. How difficult this may have been is signaled by two incidents: his friend Thomas Reinhold accompanied young Fokker to the school and correspondence home to his mother expressed severe homesickness and dissatisfaction with his living conditions. Soon, Fokker wrote home that the so-called "automotive engineering school" was a farce. Fokker stated that the school taught nothing but driving and, of the several vehicles owned by the school, only two were drivable. It was a complete waste of money. He also told his father that there was a chauffeurs' school at Zahlback near Mainz that offered a course in aircraft construction and flying, commencing in October of 1910. His father gave young Fokker his approval and Anthony Fokker entered the realm of the airplane, never to leave.

Fokker left for Mainz, Germany in October of 1910. This new venture proved to be only slightly less disastrous than his experience at the automotive school. The class diligently constructed an airplane which, sadly enough, turned out to be too heavy to fly. Their second endeavor produced an airplane that probably would have flown but did not. It seems that a baker had signed up for the class and generously donated the money to purchase an engine for the second aircraft with one condition, that he would be the first to fly the craft. When the aircraft was completed, the baker climbed into the cockpit and began taxiing down the field, and that is all he did! The airplane went straight down the field, veering neither right nor left and never left the ground. When he ran out of field, he crashed, destroying the craft beyond repair. Soon after this incident, the school closed its doors and went out of business.

Fokker then decided to build his own airplane. He had formed a friendship with a fifty-year-old, *Oberleutnant* (Lieutenant) in the German military service, *Oberleutnant* Von Daum. They formed a partnership whereby Fokker would design and construct the airframe, while Von Daum would provide the money for an engine. Still located in Mainz, Fokker and Von Daum arranged for a steel-tube framework to be manufactured in Frankfurt for the wings, while a local millwork produced wooden spars, longerons and formers for the new aircraft. Fokker hired helpers to assist in the manufacture of all of the necessary metal parts. In November of 1910, Von Daum arranged to use a vacant Zeppelin hangar at a field near Baden-Baden for final assembly. The aircraft was completed and readied for flight tests in December of 1910.

Anthony Fokker in the cockpit of the original Spin 3 in August of 1911. The aircraft used "wing warping" for control, and was steared by a control wheel. (Bowers)

Fokker named this first aircraft *Spin 1* and tested the aircraft during the latter half of December of 1910. Even though it had no ailerons or rudder, Fokker stated that the flight of the *Spin* went perfectly. Fokker kept Von Daum out of the machine as long as possible, however, his family requested that he return home for the holidays, and while he was home he received word that Von Daum had destroyed the *Spin 1*. Von Daum had fired it up immediately after Fokker left and crashed into the only tree on the field and completely destroying the aircraft. Fokker returned to Mainz immediately after being notified.

He and Von Daum teamed up with a boat-builder named Jacob Goedecker at Nieder-Walluf, for their next effort. Now living at Gonsenheim, the trio built a second *Spin* using young Fokker's plans, brains and construction skills, while Goedecker furnished space and materials. In only six weeks the aircraft was ready for testing. It first flew in May of 1911. This airplane could make turns and circle. Fokker reported that „...the plane is frightfully stable and the steering is very sensitive!" All of Fokker's monoplanes from the *Spin 1* through the E.IV employed "wing-warping" for aileron-like roll control. The pilot could introduce a roll by maneuvering the control wires attached to the top and bottom of the rear portion of the wings. By moving the control stick to the side, the rear wing surface on the side to which the stick moved was pulled up, more so at the tip. Meanwhile, the rear of the wing on the other side would be pulled down. The upward twist of the wing on the one side caused the air pressure to push the wing down. The wing on the other side was pulled down and therefore pushed up, causing the aircraft to roll along its longitudinal axis.

On 16 May 1911, Fokker received his flying certificate piloting this aircraft. Von Daum claimed ownership and *Spin 2* soon met the same fate as the *Spin 1*. Fokker purchased the remains and began construction of *Spin 3*; which was finished in some six weeks. After testing, Fokker decided to take *Spin 3* to Haarlem and demonstrate his accomplishments. After several demonstration flights, one of which was a flight over the city, he proved that he had successfully entered the world of flight.

Goedecker's relationship with Fokker was strained. Apparently a contract between them required Fokker to fly planes of Goedecker's design and act as flying instructor. About this time Fritz Cremer came on the scene. Cremer wanted to purchase an airplane and be taught to fly. Not long after, the pair entered partnership and left for Johannisthal, just outside of Berlin, to set up a flying operation. Using money contributed by Fokker's father, the *Fokker Aviatik GmbH* was recorded as a new business on 22 February 1912, just seven weeks before his twenty-second birthday. (During the war Fokker changed the name to *Fokker Flugzeugwerke GmbH*.)

July of 1912 saw the first delivery of a Fokker aeroplane to the German army at their proving grounds at Döberitz. Still sustained by his father's money, aircraft production continued, and by the end of 1912 there were eleven *Spins* in the inventory, five of which were used in his school while several had been sold to individuals. Fokker's flying school was still in dire need of income. His partner, Cremer, became chief pilot for Fokker, giving lessons to civilians and military personnel.

The German military levied a requirement on aircraft contractors that aircraft provided to the Military Air Service be easily dismantled and transportable from one location to another and then easily reassembled. Fokker demonstrated this capability and won an order for four aircraft in July of 1913. He delivered these aircraft stowed on four trucks. Six more aircraft were ordered the following month. The German air service gave these *Spin 3s* the military designation of *Militar 1* or M.1.

During this period, the military suggested that the aircraft manufacturers clustered around Johannisthal relocate. Fokker was offered thirty military trainees per year, as well as factory space and land area at a very reasonable price at Schwerin/Mecklenburg. Fokker relocated his operation there in the Autumn of 1913. Fokker moved into a small factory building approximately 50 feet by 120 feet with an annex attached. Fokker had borrowed additional money from his father and orders for aircraft continued to mount.

Fokker had an aircraft designer named Palm and assistant designers as well. New prototypes were produced. These aircraft, termed *Spin Variants*, were all monoplane designs with few similarities to the original *Spin 3*. The *Spin 3/ M.1*, was in fact the second modified *Spin 3* of 1913, which was patterned after the same aircraft produced in 1912. The variants were designated M.2, M.3, M.3a and M.4.

A reversal in Fokker's solid relationship with the military came about when the German military, after a test flight of an M.4, condemned it as unsuitable for military service.

This was not necessarily a surprise for Fokker. He had become well aware of the inadequacies of these aircraft by the Summer of 1913 after attending demonstrations of French Bleriots and Morane-Salniers. The Morane-Salnier particularly interested him. He knew that his Spins were totally overshadowed by the aerial acrobatics performed in these flying machines.

The condemnation of the M.4 led Fokker to obtain a damaged Morane-Salnier (probably a type H) and rebuild it at Schwerin. He and his associates flew the aircraft and recognized its potential. He fired his chief designer Palm, and appointed Palm's assistant, Martin Kreutzer, to his position. Fokker, Kreutzer, and others on his staff went to work on a completely new design.

A line up of Fokker aircraft used by the students at the Fokker Flying School at Johannisthal, just outside of Berlin. (Bowers)

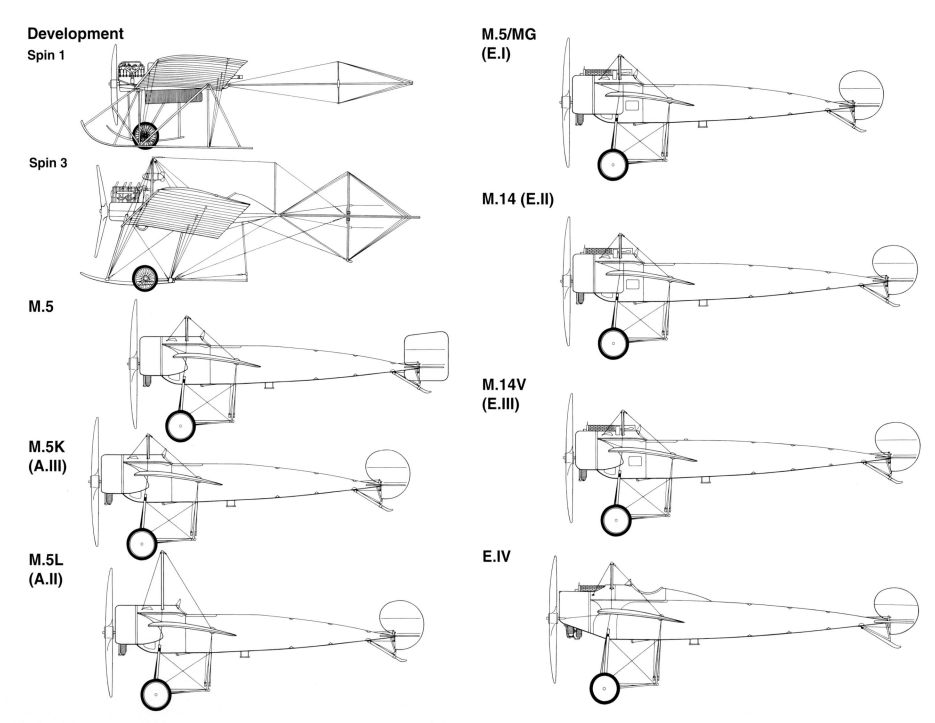

The Fokker M.5 (Eindecker I)

The small monoplane scout went into a shallow dive, pulled up into a climb, fell over onto its back, flew inverted, then dove straight at the ground. At what seemed to be the last minute, it leveled out and flew across the field, only a few hundred meters off the ground. Anthony Fokker, demonstrating his newest monoplane in May of 1914, had just completed a loop. Among his audience was General Von Falkenhayn, the German minister of war, accompanied by his advisors. Fokker's demonstration must certainly have impressed the High Command, because he was called in a short time later and given an order for a two-seat monoplane and a biplane. The biplane may have been the Fokker M.7 or Fokker B. The Fokker B had two cockpits with a fuselage and empennage identical to the monoplane M.5. The lower wing appears identical to that of the M.5, with the top wing simply matching the shape and span of the lower wing. The lower wing was rooted at the bottom of the fuselage. Initially the monoplane M.5s were classified as "A" types. The military used "A" to designate monoplane scouts, while "B" was used to designate biplane observer aircraft. When machine guns were mounted, a "C" designated a armed biplane observation aircraft, a "D" designated a biplane fighter and a "E" designated a monoplane armed scout.

A number of historians, including J. M. Bruce refer to the M.5 monoplane designed by Fokker and his staff as a "carbon copy" of the French Morane-Salnier (H). Others, including Henri Hegener are less harsh, referring to "similarities" that existed between the Morane-Salnier and the Fokker M.5.

A direct comparison reveals a great similarity between a Morane-Salnier Type H and the Fokker M.5 prototype. Construction techniques of the two aircraft, however, were vastly different. The manufacturer of the Morane-Salnier built most of the aircraft from wood, a practice common in the aircraft industry at that time. Metal cowlings and a large amount of wire bracing were standard. Fokker constructed all but the wings of the M.5 prototype from wire-braced, steel tubing with welded joints. The M.5's landing gear was much sturdier than that of the Morane-Salnier, with a wider wheel track.

The design team at Fokker built the prototype M.5 with a rectangular, balanced rudder somewhat similar to that of the Morane-Salnier. This rectangular rudder did not provide control response that Fokker wanted, so they installed a "comma" shaped rudder with greater height and a larger area. This comma-shaped Fokker rudder became a characteristic signature

The German military purchased a number of *Spin 3* type aircraft under the designation M.1. This aircraft was registered as A38/13, with the A indicating the aircraft was a monoplane scout. (Bowers)

This *Spin 3* variant (designated M.4) was found to be unacceptable by the German military and rejected. This led Fokker to design the M.5 which evolved into the Fokker E.1. (Grosz)

for Fokker Aircraft until the Fokker D-VII was introduced late in 1917.

The Fokker Company had not incorporated a rotary engine in any of the aircraft it produced until the M.5 prototype, although rotaries were popular in French designs and used in some British designs. In fact, the rotary engine used in the prototype M.5 was the Gnome, a 50 hp French rotary that had been salvaged from the rebuilt Morane-Salnier at the Fokker factory. A rotary engine has the crankshaft fixed to the aircraft motor-mounts. The propeller is bolted to the engine casing to which the cylinders are attached. When the engine is started, the engine rotates, and when each piston fires, the thrust of the power stroke is against the fixed, cammed crankshaft in such a way that a force is exerted in the direction of engine rotation.

The Fokker design team did not intend to obtain its rotary engines from French manufacturers. The German firm, **Oberursel Moteren Gesellschaft**, had established a licensing agreement before the war with the French manufacturer that allowed Gnome engines to be produced in Germany. The first shipment of 80 hp, seven-cylinder rotaries to the Fokker factory was delayed, requiring the installation of the 50 hp Gnome rotary in the first prototype. It is unknown if the first prototype was specifically designed to accommodate the smaller 50 hp engine for testing or had been planned for the larger engine from the start. The second M.5 prototype produced, however, was considerably different than the first in that it had a longer

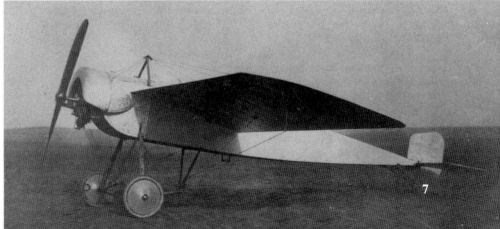

The first M.5 prototype had a retangular shaped rudder that proved to be unsatisfactory and was replaced by a comma shaped rudder. This aircraft was designated by the German military as the M.5K. (Grosz)

Morane-Salnier Type L

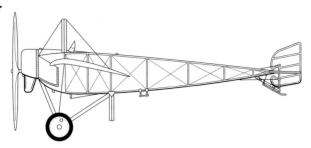

Fokker M.5

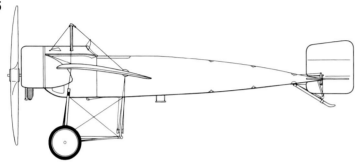

The cutouts on the wings of this M.5 (A) enabled the pilot to see below the aircraft for visual reconnaissance. These were later replaced by panels in the fuselage. (Grosz)

alry was rapid reconnaissance.

The M.5s (A.IIIs) had a single cockpit, however, this cockpit was sufficiently elongated so that an observer or passenger could sit behind the pilot. At that time the pilot sat on a bench-type seat, lengthened to accommodate a second person. Several methods were employed to aid the pilot in observing ground activity. Rearward sections of the wing next to the fuselage were left uncovered so that the ground could be seen when looking over the sides of the aircraft. Later, small openings (introduced on the E.II) were provided under the wings near the bottom of the fuselage that had accordion-like fabric coverings that enabled the pilot to open or close them as needed. Still later in the evolution of the E series, metal-covered, rectangular viewports were installed directly under the wings and additional view ports were installed

Prototype Development

wingspan and fuselage, with a much higher pylon to support and control the longer wings. This longer span provided more lift for the larger and heavier 80 hp engine. (J. M. Bruce states that a 70 hp Gnome was installed in this prototype and then replaced by the 80 hp Oberursel rotary.)

Fokker replaced the 50 hp Gnome engine in the first M.5 prototype with a 80 hp Oberursel once the engine was available and the two prototypes were then turned over to the German military to determine which version they preferred. The military designated the shorter-winged M.5, with an approximately 27.8 foot (8.5 meter) wing span, as the M.5K. The K stood for *kurz*, German for short. They designated the M.5, with the 31.5 foot (9.6 meter) span, as the M.5L, the L signifying *lang*, or "long." The M.5K proved to be slightly faster, however, while the M.5L was more responsive. In the end, the military chose the M.5L.

There are different references for the height of the two aircraft, Hegener lists the overall height of both the K and L as 9.5 feet (2.9 meters), while J. M. Bruce indicates that the height of the M.5 as 10.1 feet (3.1 meters). This difference probably stems from the height being measured separately for the K and the L.

The First World War began on 1 August 1914 when The Austro-Hungarian Empire invaded Russia. A short time later, Germany invaded Belgium and France. The German military had been mobilizing for war during most of 1914, and in July had ordered M.5s off the factory floor as fast at they were being built. These Fokker M.5s were assigned to observer groups. The military had established Flying Sections (*Flieger Abteilungen*) consisting mainly of two seat observation biplanes that supported the troops and artillery involved in ground action. Fokker and Pfalz provided small contingents of monoplanes to these flying sections, and these aircraft were sometimes referred to as "Cavalry Scouts," since one of the missions of the cav-

M.5K

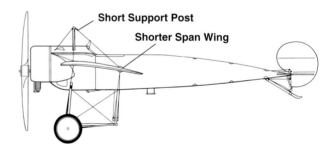

M.5L

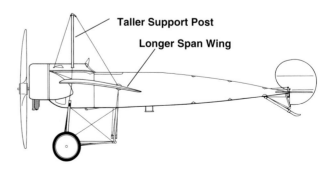

(Above and Right) Anthony Fokker poses with a Fokker M.5L (A) unarmed reconnaissance scout on the factory grounds. The M.5L had a longer wing span and much taller support strut in front of the cockpit than the M.5K. (Bowers)

behind the front strut supports of the landing gear, in the bottom of the fuselage. This double port could be opened much the same as bomb bay doors, so the pilot could observe the ground from between his legs. The Fokker scouts were popular with German pilots, Henri Hegener states that in December of 1914, Oswald Boelcke, later to become one of Germany's top aces, wrote, *"The Fokker is my best Christmas present in which I take childish pleasure."* At the time, Boelcke flew with *Flieger Abteilung 2*.

Regardless of the German successes on the ground (their troops had nearly reached the outskirts of Paris in just thirty-five days), their successes in the air were nonexistent and their observer aircraft were frequently shot down. From the beginning of the war, French, German and British aircraft on observation sorties had fired at each other with pistols and rifles, and had even tried to drop hand-bombs on one another. This aerial warfare escalated when both sides mounted machine guns on their planes for the observers. Limited by an extremely small field of fire, very special combat tactics evolved. Adversaries attempted to fly alongside or under one another so that the observer's gun could be effectively brought to bear. The role of observer aircraft did not change, however, their primary mission remained to fly over enemy troop positions and report back to their commanders any tactical changes that occurred.

The '08/14 Parabellum machine gun was generally used by observers on German aircraft. This machine gun was designed by Kurt Heinemann, who was given the task of lightening and improving the standard '08 Maxim machine gun used by the troops, and manufactured under license by the Spandau Armory. The French and British generally used the standard Lewis machine gun for their observers. Then a definite turn of events in aerial combat occurred early in 1915, when the French began to mount forward-firing machine guns on their single-seat scouts. These machine guns, nearly inaccessible to the pilot, fired over or to the side of the propeller arc. German losses in the air increased significantly. Several French pilots were given the rank of "Ace" (five victories) before winter was over. The German military, with no counter-measure at hand, became very concerned over the Allied successes.

One of these French pilots, Roland Garros, became famous. Some authors have stated that, on his own, Garros fitted "bullet deflectors" to the propeller of his Morane-Salnier type L scout. He then had a French Hotchkiss machine gun fitted to his cowl. When the gun fired, any rounds hitting the propeller blades would be deflected to either side. In all probability;

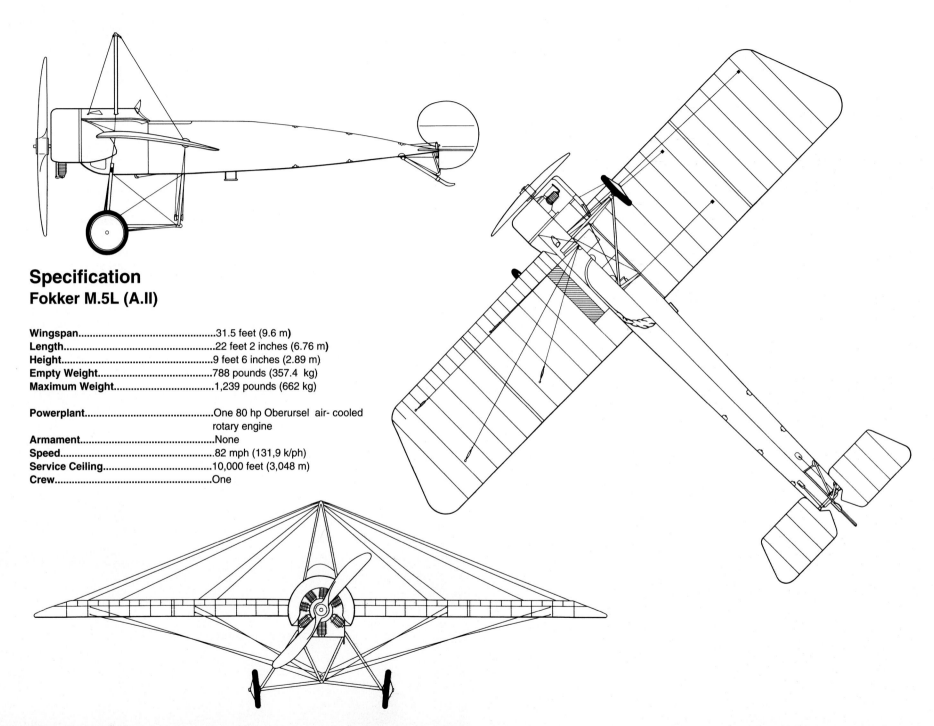

Specification
Fokker M.5L (A.II)

Wingspan...31.5 feet (9.6 m)
Length..22 feet 2 inches (6.76 m)
Height..9 feet 6 inches (2.89 m)
Empty Weight..788 pounds (357.4 kg)
Maximum Weight..................................1,239 pounds (662 kg)

Powerplant...One 80 hp Oberursel air-cooled rotary engine
Armament..None
Speed..82 mph (131,9 k/ph)
Service Ceiling.....................................10,000 feet (3,048 m)
Crew..One

A line-up of Fokker M.5Ls (As) at the Fokker factory. The aircraft have a mix of shoulder and high mounted wings, while those at the end of the line-up do not have their wings mounted. (Grosz)

however, the Morane-Salnier factory had a hand in the design since Garros did visit Salnier, president of the Morane-Salnier company, for help and the device was fitted not only to his but other pilots' aircraft as well. Garros achieved the rank of Ace by April of 1915 using the machine gun deflector arrangement This is not surprising; he had flown racing aircraft before he entered the French Air Service and was an accomplished pilot.

On 16 April 1915, a stroke of bad luck for the French and good luck for the Germans occurred when Garros had to land his plane behind German lines. Apparently engine trouble had brought him down. Before he could destroy the aircraft, German troops captured him and the airplane intact. The aircraft was soon in the hands of the German High Command. It did

Lieutenant Von Loessl (left) was assigned to *Flieger Abteilung 21*. The M.5L (A.II) these pilots are leaning on is an early aircraft with no armament and what appears to he an altimeter suspended between the wing support pylons. (Bowers)

A Fokker M.5L (A.II) in flight. The seam running along the fuselage underside suggests that the fabric covering was applied like a sleeve. (Grosz)

not take long for the Germans to try out this new approach to aircraft armament. When they did try it (perhaps on a Pfalz scout) the test pilot promptly shot his propeller to pieces. Lack of success for the Germans can be attributed to the difference in ammunition used. The French used copper-jacketed slugs as standard issue; whereas the Germans used steel-jacketed slugs. The softer French ammunition easily deformed and ricocheted off the steel wedges on Garros' propeller. The steel-jacketed ammunition used by the Germans resisted deformation to a much greater extent and caused the wedges, and subsequently the propeller, to shatter.

Henri Hegener has stated that the German Air Ministry called Anthony Fokker into Berlin and presented him with the problem of how to fire a machine gun through the propeller arc.

The seven cylinder Oberursel rotary was a license built Gnome engine. The cowling was supported by the framework visible just below the propeller blades. (Grosz)

This M.5L (03.51) was built for the Austro-Hungarian Air Force and was the first to enter service. It was later armed with a machine gun and served with Flik 4.

The Fokker M.5/MG was the prototype for the armed M.5 (A) aircraft. The aircraft, work number 216, was armed with a 7.92 ᴍᴍ 08/14 Parabellum machine gun. (Grosz)

Within days, Fokker returned to Berlin from Schwerin with his solution: an M.5K on a trailer behind his automobile with a Parabellum machine gun and synchronizer installed on it. J.M. Bruce states that Fokker had little to do with the design of the synchronizer. This is similar to stating that a project leader has little to do with the project.

A German patent existed at this time for a synchronizer or interrupter system. Franz Schneider of the German LVG Aircraft Works had obtained a patent in 1913 on an interrupter mechanism. At the time, the design existed only on paper and had never been tried. The patent described a system whereby the propeller rotation interfered with the actual operation of the machine gun.

In the winter of 1913, Raymond Salnier developed a synchronizer past the "paper" stage. It

This was a common method of delivery for Fokker fighters. The aircraft was partially disassembled and towed to the delivery location, often by Fokker himself. This M.5K has the propeller stowed between the landing gear wheels. (Fokker)

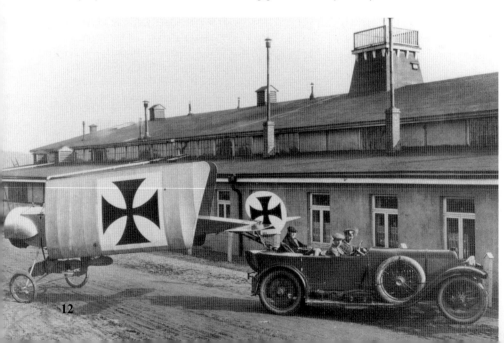

Armament Development

Fokker M.5K (A.III) — No Armament

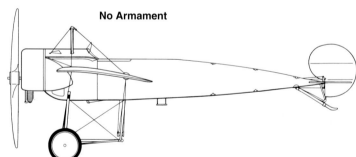

Fokker E.I — Synchronized Machine Gun, Either 08/14 Parabellum or 08 Standard

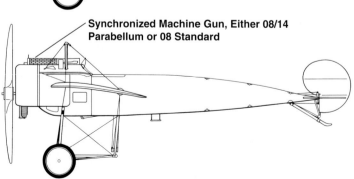

Oberleutnant Max Immelmann poses in front of his Fokker E.I. The wartime censors have retouched the original photo, removing the details of the engine and gun. (Bowers)

worked; but not consistently. Evidently the inconsistency in the discharge rate of ammunition (hang-fires) defeated it. This problem could also have affected the Fokker device, but apparently the superior quality of German ammunition greatly reduced the occurrence and duration of hang-fires.

This Fokker M.5/MG (E.I) was assigned to a front line unit. The gun button and magneto cut-out spade are visible on the control stick. The weapon was a 7.92MM 08/14 Parabellum machine gun. (Jarrett)

Oberleutnant Von Althaus in the cockpit of the Fokker E.I flown by Oswald Boelcke. (Bowers)

The controversy among historians about the Fokker synchronizer does not lessen the fact that the Fokker Company produced a working device. Theirs was a deceptively simple solution to the synchronization problem. The Fokker Company designed a synchronizer that used the engine rotation to fire the machine gun! The Maxim-designed, short-recoil machine gun (both the parabellum '08/14 and the '08 standard) had a trigger sear that moved backward and upward when the gun fired. A firing sequence began when the trigger bar, in the lower portion of the gun receiver, was moved backward by depressing the trigger. This caused a raised portion or lip on the trigger bar to pull the trigger sear backward, firing the gun. As long as the trigger bar was in this position each time the trigger sear moved back to its cocked position, it hit the lip of the trigger bar and fired again. This cycle continued until pressure was released on the trigger bar, thus moving the lip out of position so that the trigger sear would not make contact with it when completing the cocking cycle.

The Fokker team did two things: they hooked the trigger bar of the machine gun to a connecting rod running out the front of the machine gun, through an L-type rocker, to a cam-follower rigidly attached to the engine firewall at the rear of the rotating engine. A one or two-lobe firing cam was attached to the rearmost portion of the engine, just behind the engine spark-ring. The lobe of the cam was positioned between the propeller blades such that the trigger bar moved into firing position in the machine gun only when the propeller blades were not aligned with the barrel of the machine gun.

Secondly, they separated the lip of the trigger bar from the portion "jiggled" by the engine cam, with a moveable, intermediate section. Pressing the trigger on the stick, via a bowden cable, dropped the moveable section into position between the portion of the trigger bar with

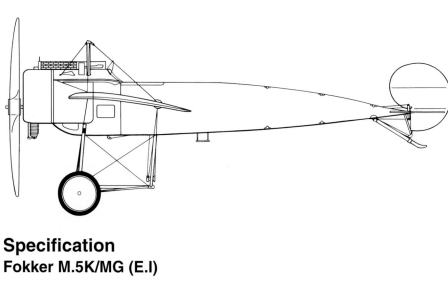

Specification
Fokker M.5K/MG (E.I)

Wingspan...28 feet (8.53m)
Length..22 feet 2 inches (6.76 m)
Height...9 feet 6 inches (2.89 m)
Empty Weight......................................788 pounds (357.4 kg)
Maximum Weight................................1,239 pounds (662 kg)

Powerplant..One 80 hp Oberursel air-cooled rotary engine
Armament...One forward firing 7.92mm machine gun
Speed...82 mph (131,9 k/ph)
Service Ceiling....................................10,000 feet (3,048 m)
Crew...One

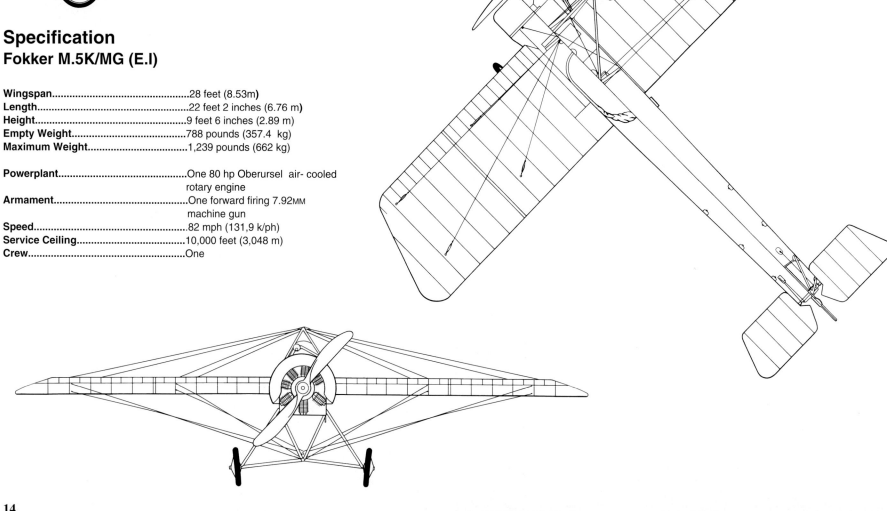

This M.5/MG (E.I) was purchased by the Austro-Hungarian Navy and given the side number A6. (Fokker)

Oberleutant Hesse in the cockpit of his Fokker E.I This is an early aircraft with no fuselage insignia or serial numbers. (Bowers)

the firing lip and the portion of the trigger bar being jiggled by the cam-follower. Spring-action kept the three pieces of the trigger bar married to one another until release of the trigger allowed the moveable section to move out of position, disconnecting the firing lip from the cam-action of the engine. In this way regardless of engine rpm, the gun could only fire when the propeller was out of the way. The engine essentially pulled the trigger of the machine gun.

Deterioration of the rate-of-fire of the machine gun was minimal. Henri Hegener states that the firing rate for the Maxim, both the '08 and '08/14, was 600 rounds per minute. A direct quote from Hiram Maxim, circa 1900, gives a firing rate of around 450 rounds per minute. A rough calculation yields a reduced firing rate from 600 to 526 rounds per minute, or a twelve percent decrease for a single-lobe firing cam. With two lobes, the reduction is only about 1.5 percent.

A discussion with the E.III Replica Project Leader at the San Diego Aerospace Museum revealed that a mechanical device was installed that relieved the constant vibration of the trig-

Fokker work number 206 was a very early M.5/MG (E.I) serial 8/15. It may have been the first to mount a standard 7.92MM 1908 Maxim LMG. (Fokker)

ger bar caused by the rotation of the engine. The Author has not seen any description of this capability in any historical literature. The death of one of Germany's most famous Aces, Max Immelmann, is sometimes attributed to the failure of the synchronizer, which in turn caused his machine gun to demolish his propeller, sending him crashing to his death.

Thirty Fokker As were ordered by the German military to be armed with machine guns under the designation E.Is (Fokker M.5L/MG)

The military assigned several of the armed E.Is to Döberitz for pilot training. The remainder were assigned to airfields close to the front which were to be flown by experienced pilots. Lieutenant Kurt Wintgens was assigned E.I 2/15, in the Flanders area. Here, Wintgens downed his first enemy aircraft on 1 July 1915. In mid-July of 1915, Fokker, accompanying deliveries of the newly armed E-Is, arrived at Douai, France where two M.5L/MGs, E.Is 1/15 and 3/15, had been assigned. The airfield was the station for *Flieger Abteilung 62*. These two aircraft were assigned to Lieutenants Max Immelmann and Oswalde Boelcke. Max Immelmann had been assigned to *Flieger Abteilung 62* when it formed at Döberitz in May, of 1915. Prior to that time he had flown for artillery spotters (observers) from the German base at Rethel in the Aisne valley.

The 62nd Flying Section was primarily outfitted with LVG two-seater biplanes The LVG was one of the first German aircraft to mount a Parabellum machine gun for use by the observer. This air group flew observation support for the German forces fighting in that area. Apparently, tenure in the Air Service and competence were the reasons that Immelmann was assigned to one of the first E.Is armed with a machine gun. Oswald Boelcke, assigned to *Flieger Abteilung 62* at the same time as Immelmann, had also established himself with the group as a competent, experienced pilot.

Both Immelmann, later known as "The Eagle of Lille," and Boelcke distinguished themselves above and beyond the call of duty. Immelmann and Boelcke earned the highest honor for German airmen, the *"Ordre Pour le Merite"* or the "Blue Max," given for eight victories in the air. Immelmann went on to down eighteen allied aircraft before being killed on 18 June 1916, in a crash caused by major aircraft structural failure. At the time of his death he was fly-

An M.5K/MG (E.I) at the factory with a 7.92MM 08/14 Parabellum marchine gun mounted on the cowling. (Grosz)

ing an E.III. Later, on 28 October 1916, Oswald Boelcke suffered a collision with one of his newer pilots (he was then in command of **Jagdstaffel 2** and had forty victories). He crashed near the Somme and was killed. Neither pilot had died as a victim of Allied airmen.

A truck with an M.5K.MG (E.I) in tow ready for delivery to a front line unit. This was a common method of delivery for new production aircraft. (Grosz)

The longer "bathtub" cockpit of the M.5K/MG (E.1) is quite evident. This aircraft was serialed 46/15. (Bosers)

Armed and unarmed, Fokker E.Is saw service in the German Military Air Service until 1917. They were phased out of combat and used behind the lines and at military flying schools until they were completely displaced by more advanced aircraft According to J. M. Bruce, approximately sixty-five E.Is and E.IIs were built, no distinction made between the two types, while according to P.M. Grosz, sixty-eight E.Is and E.IIs were accepted for delivery by the German Air Service. The overall output of the Fokker factory went four ways throughout the war: the German Army, the German Navy, the Austro-Hungarian Empire and Turkey, with by far the most aircraft going to the German Army.

Lieutenant Von Althaus in the cockpit of an armed M.5L (E.I). This aircraft was reportedly also flown by Oswald Boelcke. (Bowers)

Fokker M.14 (Eindecker II)

Historian P.M. Grosz estimates that only twenty-four to thirty-six Fokker E.IIs were actually built! Even more astounding, just two months after the first victories with the E.I, the Fokker E.III appeared in service! There are very good reasons for this rapid introduction of new variants of the original Fokker monoplane. First and foremost, the role of this aircraft changed from that of observation and reconnaissance to that of offensive fighter. The phase-in of the E.II began when the 100 hp nine-cylinder Oberursel rotary engine became available. The nearly thirty-percent increase in engine weight required that the fuselage be lengthened to maintain the aerodynamic C/G (center of gravity). The diameter of the cowling was increased about three to 4 inches (eight to ten centimeters) to accommodate the larger engine. The fuselage formers and longerons may have been made heavier, but the Author.has no indication that this actually took place.

The wingspan remained roughly the same as that of the M.5L. J. M. Bruce states that the span for the E.II was shortened from 31.5 feet (9.6 meters-M.5L) to 29.5 feet (9 meters) and lists the wing area as 45.9 square feet (14 square meters). Another source, Heiri Hegener states the wing span of the E.II was 31.98 feet (9.75 meters).and also states that the fuselage was lengthened about 1.64 feet (.5 meter) so that the overall length of the Fokker E.II was about 23.9 feet (7.3 meters).

Apparently the Fokker factory was in trouble during June of 1915. The "monkey-wrench in the works" was the machine gun. Fokker had the E.I in service, and the German military was taking all of these he could produce. Kreutzer and the design team had a modified version of the E.I, the E.II, just off the drawing boards and going into production. This variant would accommodate the bigger and supposedly more reliable 100 hp Oberursel rotary engine (engine problems had plagued the E.I from the start). With no former knowledge of what changes the

A M.14 (E.II 69/15) parked on wooden planking at the factory field. The small lettering on the leather patches along the fuselage says "Lift Here" in German. The aircraft has a modified cowling, indicating that it was probably undergoing conversion to E. III standards. (Bowers)

Fuselage Development

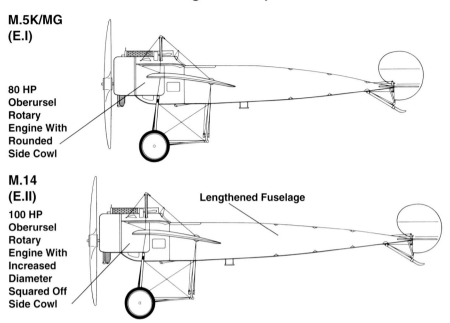

M.5K/MG (E.I)

80 HP Oberursel Rotary Engine With Rounded Side Cowl

M.14 (E.II)

100 HP Oberursel Rotary Engine With Increased Diameter Squared Off Side Cowl

Lengthened Fuselage

A Fokker M.14 (E.II) at the factory, ready for rail shipment to the front. The sign states that reassembly instructions are on the seat and that they are to be followed precisely. (Grosz)

A Fokker E.II (M.14) is readied for rollout from the Fokker factory. It is believed that this may have been the last E. II built. (Bowers)

machine gun and the ammunition would require, Fokker and his team had been rolling along with scheduled improvements as they were required, redesigning for the 9-cylinder, 100-hp engine. (Note: if Fokker had already developed a synchronizer just waiting to be installed, as

German ground crewmen make adjustment to a Fokker E. II (M.14), serial 35/15 assigned to Flieger Abteilung 14 on the Western Front. (Bowers)

Hauptmann Von Buttlar inspects Fokker E.II 25/15. Von Buttlar had earlier conducted extensive tests of the E.I (serial 2/15) delivered by Lieutenant Wintgens to Flanders. (Bowers)

some have suggested, the E.III and not the E.II would have been next into production).

On the E.I and E.II, the gasoline and oil tanks were nestled under the top cowling and inside the fuselage, forward of the pilot. Installation of a machine gun directly over these tanks required that space under the machine gun be allocated for ammunition storage. The E.II was not designed for this. Most of the photographs of the right front side of the fuselage of the E.Is and E.IIs show several different ways that ammunition belts were strung up from the area below the pilot's legs and under the rudder cables. The empty belts were channeled down the left side to an area not in the way of the pilot's legs or the live ammunition.

The reason that the E.III came out so soon after the E.II was really quite simple: a large gasoline tank, about one hundred liters, was installed in the E.II directly behind the pilot. This made room under the cowl for an ammunition magazine and spent ammunition-belt box in front of the pilot. The cockpit was designed to be that of a true *Einsitzer* (single seater) and

This modified Fokker E.II had a fully circular cowling and an oil sump for collecting used castrol. (Bowers)

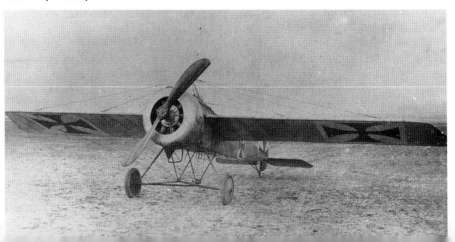

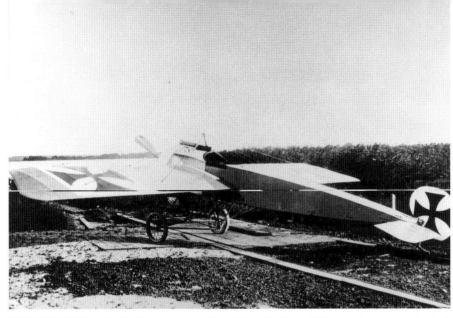

M.14 (E.II) 69/15 parked in the open near the Fokker factory. This may have been the last Fokker E.II to be produced. (R. R. Martin)

Lieutenant Ernest Freiherr von Althaus, of *Flieger Abteilung 23* poses along side a Fokker M.14 (E.II) fitted with an adjustable headrest. (Nowarra)

Corporal Boehme in the cockpit of a Fokker M.14 (E.II) serial 43/15. The wooden propeller has a German cross in the blade. (Bowers)

Lieutenant Kurt Wintgens in flight with his Fokker M.14 (E.II) There is an adjustable headrest visible just behind the pilots head. (Bowers)

This Fokker M.14 E.II 35/15 was being wheeled away by its ground crew for maintenance at a forward air field. (Bowers)

This Fokker E.II has no windshield fitted. It was armed with a 7.92MM 08/14 Parabellum machine gun and had an adjustable head rest for the pilot. (Fokker)

Fokker E.M.14 (E.II) 7/15 flipped over onto its back after a crash landing at a German front line airfield. The aircraft has broken its back just behind the cockpit and the tail group had been broken at the rudder hinge line.

Tony Fokker at the controls of what appears to be Work Number 206, which was personalized for Fokker with his name on the fuselage in Black. (Grosz)

only a small remnant of the original gasoline tank was left intact under the cowl. This tank now held about twenty-two liters for reserve. This arrangement seemed to work so well that the few E.IIs in service were brought back to the factory to have these changes retrofitted, causing a great deal of confusion regarding the reclassification of a number of E.IIs as E.IIIs. This also supports the conclusion that there was very little difference between the E.II and the E.III other than the gasoline tank and ammunition storage arrangements. In fact, there was no change in the military designation of M.14, although the Type-designation changed from E.II to E.III.

(Right) Although this aircraft carries a E.I designation, it has been outfitted with conversion parts used to convert E. IIs to E.III standards. The work number, 309, is a much later number than used on any E.I and the aircraft is armed with a 08 Standard machine gun. (Bowers)

The Fokker M.14V (Eindecker E.III)

Correspondence to the Author from H.A. Somberg, Fokker VFW International, Holland, in October of 1973, revealed that 268 Fokker E.IIIs were built, according to factory records in his possession. Historian P. M. Grosz also quotes this exact number along with thirty-two additional units being manufactured for other than the German Army.

The Fokker E.III achieved operational status by September 1915, just a few months after the Fokker E.II entered service. There was very little difference between the E.II and E.III, other than adding a new gasoline tank behind the pilot and eliminating most of the gasoline tank under the front cowl, improved ammunition storage for the machine gun and storage for spent belts. During the production life of the E.III, about twelve months, the wing span of the E.III increased to about 32.8 feet (10 meters) from the 31.4 feet (9.6 meters) of the E.II. It is uncertain if this change in wingspan occurred gradually or as a specific modification.

The success of the E.III in the air as a fighter is evidenced by the uproar in England over the significant losses in Allied aircraft that this "menace of the air" was causing. Newspapers claimed the Fokker Es were the "Fokker Scourge." The British parliament exerted great pressure on British aircraft manufacturers to come up with a way of ending the high casualty rates suffered by both the British and French squadrons.

Not only were the Fokkers deployed to an ever-increasing number of flying sections along the fronts, but German air tactics had matured to the point that KEK (*Kampf Einsitzer Kommando*) groups began forming. The flying officer for the German Fifth Army initiated the KEK concept where fighter (Eindecker) aircraft from several *Flieger Abteilungen* could be called up to attack Allied aircraft over the front during the battle of the Somme River at Verdun. Initially **KEK Nord** (north), **KEK Sud** (south) and **KEK 3** led by Boelcke, were

Anthony Fokker poses beside a Fokker Eindecker. The E.III was the most widely produced variant of the Eindecker series with over 2160 being produced. (Bowers)

Development

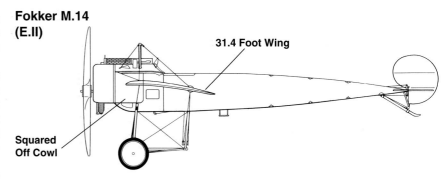

Fokker M.14 (E.II) — 31.4 Foot Wing, Squared Off Cowl

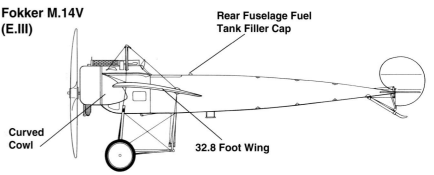

Fokker M.14V (E.III) — Rear Fuselage Fuel Tank Filler Cap, Curved Cowl, 32.8 Foot Wing

formed. These were followed by at least three additional KEKs a short time later.

A second organizational step was taken by the German Air Service a short time later that totally changed air force organizations. Oswald Boelcke had been sent to Turkey at the

The Bavarian Kronprinz discusses the merits of a Fokker E.III with Tony Fokker. This aircraft has an oil sump under the lower cowling. (Author)

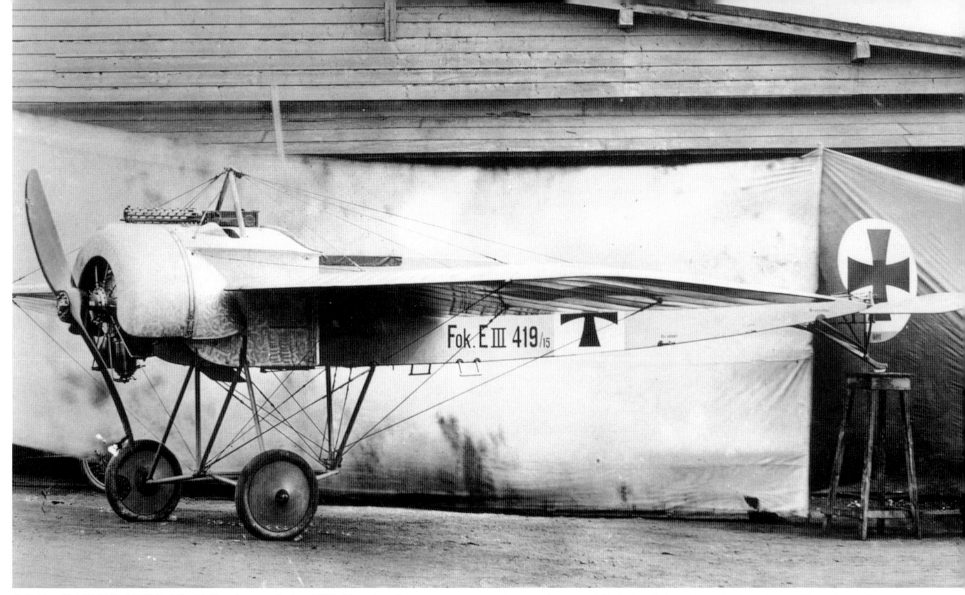

A Fokker M.14V (E.III, 419/15) at the Fokker factory during 1915. The aircraft was powered by a 100 hp Oberursel rotary engine and armed with a single 7.92MM 08 machine gun. (Fokker)

request of German royalty to keep him from being killed, since he was the highest ranking Ace in the German Air Service and made an excellent morale booster for the German public. The German high command called him back from Turkey for a series of conferences to formulate the best methods for the organization of fighter aircraft. The concept of the *Jagdstaffeln* was born from these conferences.

Jagdstaffeln, or *Jastas* (hunting groups), were formed with fighters whose only role was to attack, in formation, any Allied aircraft found over the front lines. These *Jastas* flew just behind the German lines and attacked any Allied aircraft seen flying over or near the lines. It should be noted that Oswald Boelcke had been selected to form *Jasta 2*; which, after his death, would be commanded by the "Red Baron," Baron Manfred Von Richthofen (*Jasta 1* existed only on paper).

Both the French and the British came up with answers to the Fokker menace, neither involving synchronized machine guns. Manufactured by the French and first put into service by the British, was the Nieuport 11. This aircraft went into service following the Nieuport 10, a two-seater observation craft. There were vast differences, however, between these aircraft. The

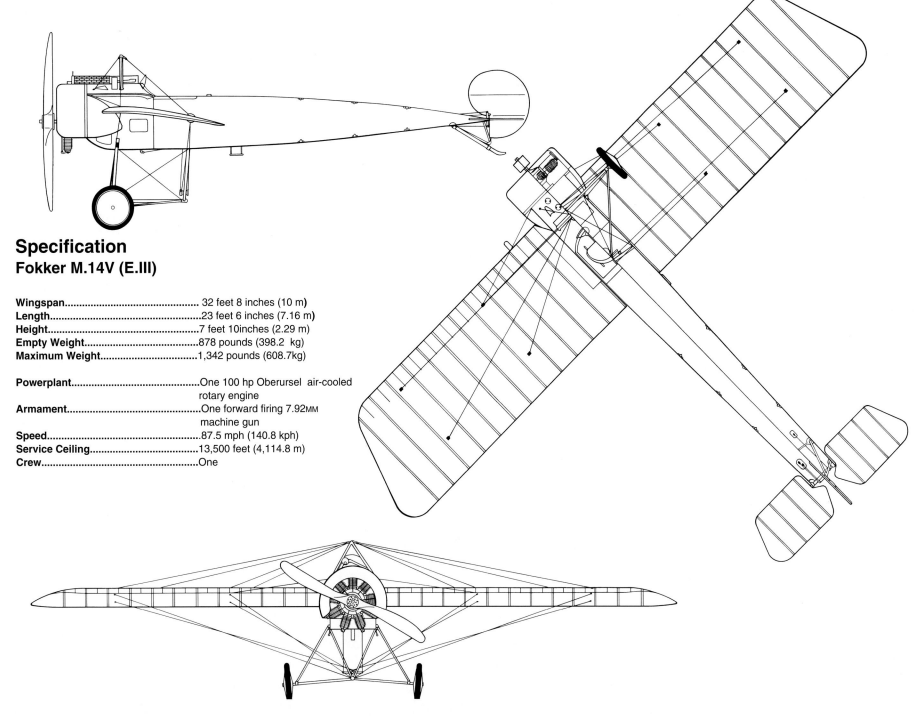

Specification
Fokker M.14V (E.III)

Wingspan	32 feet 8 inches (10 m)
Length	23 feet 6 inches (7.16 m)
Height	7 feet 10 inches (2.29 m)
Empty Weight	878 pounds (398.2 kg)
Maximum Weight	1,342 pounds (608.7kg)
Powerplant	One 100 hp Oberursel air-cooled rotary engine
Armament	One forward firing 7.92mm machine gun
Speed	87.5 mph (140.8 kph)
Service Ceiling	13,500 feet (4,114.8 m)
Crew	One

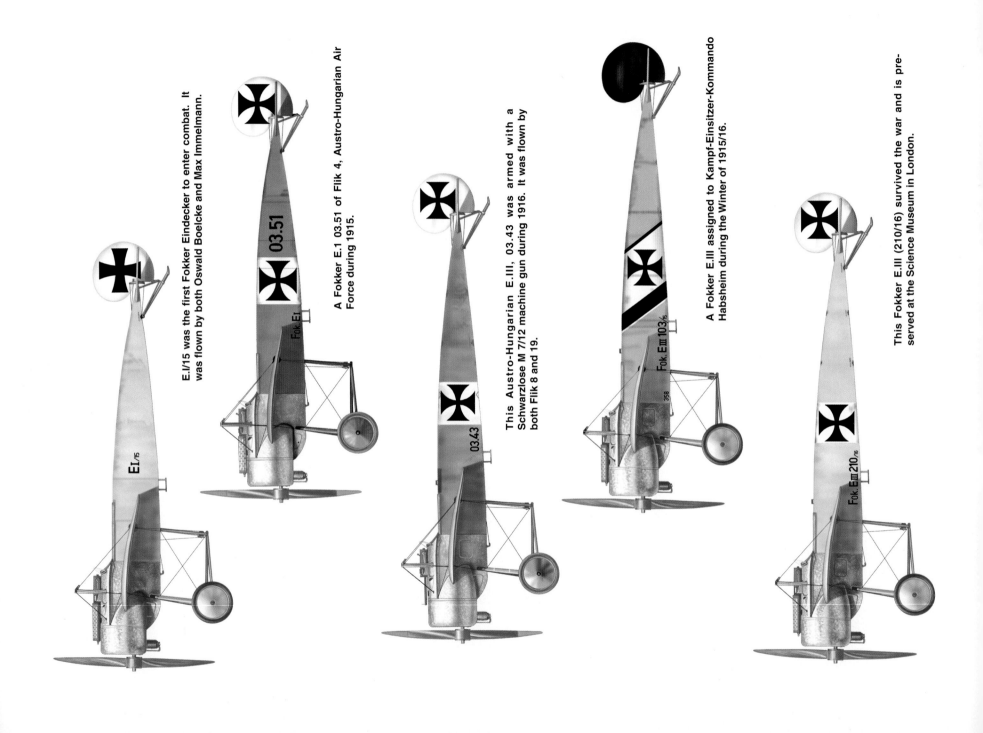

E.I/15 was the first Fokker Eindecker to enter combat. It was flown by both Oswald Boelcke and Max Immelmann.

A Fokker E.1 03.51 of Flik 4, Austro-Hungarian Air Force during 1915.

This Austro-Hungarian E.III, 03.43 was armed with a Schwarzlose M 7/12 machine gun during 1916. It was flown by both Flik 8 and 19.

A Fokker E.III assigned to Kampf-Einsitzer-Kommando Habsheim during the Winter of 1915/16.

This Fokker E.III (210/16) survived the war and is preserved at the Science Museum in London.

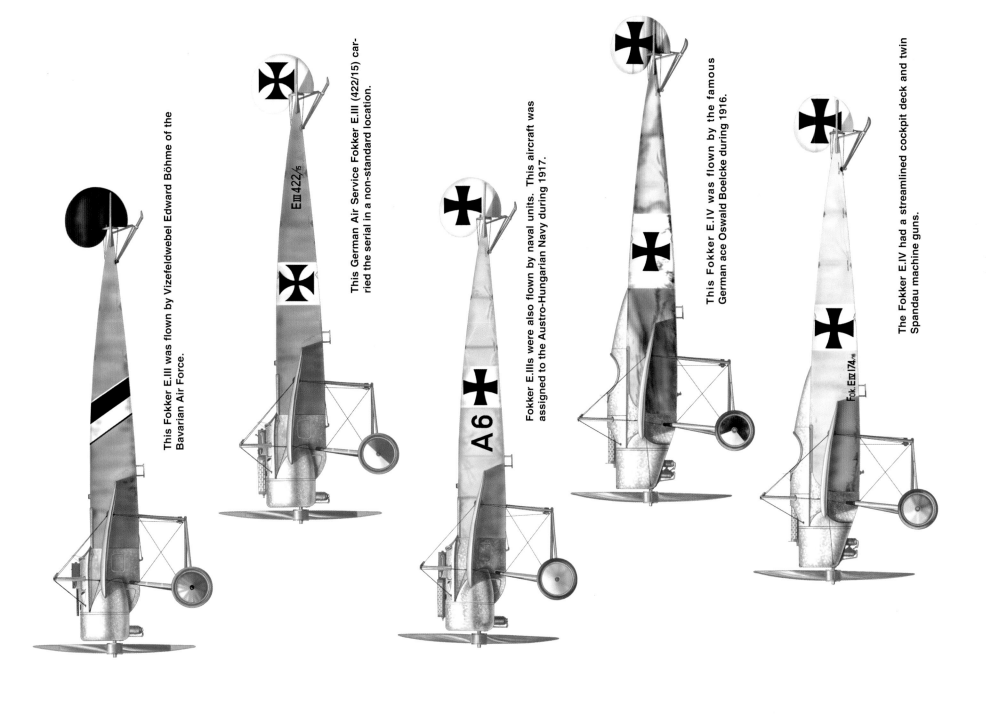

This Fokker E.III was flown by Vizefeldwebel Edward Böhme of the Bavarian Air Force.

This German Air Service Fokker E.III (422/15) carried the serial in a non-standard location.

Fokker E.IIIs were also flown by naval units. This aircraft was assigned to the Austro-Hungarian Navy during 1917.

This Fokker E.IV was flown by the famous German ace Oswald Boelcke during 1916.

The Fokker E.IV had a streamlined cockpit deck and twin Spandau machine guns.

Eindecker Armament

08/14 7.92MM Maxim Parabellum Machine Gun

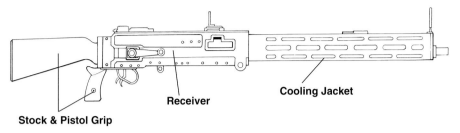

08 7.92MM Standard Maxim (Lightened)

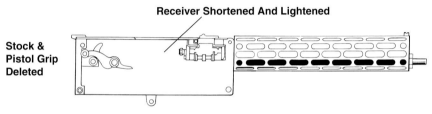

small Nieuport 11, referred to as *Bebe* or "one-and-one-half plane," was a small, single-seat biplane (13 square meters of wing area for both wings) which was very fast and highly maneuverable. In fact, the aircraft had been originally designed before the war for air racing in France. It mounted a single Lewis machine gun on the top of the upper wing, firing over

This Fokker M.14V (E.III) 401/15 was flown by Lieutenant Von Zastrow. The aircraft carried no compass and there was an unknown device attached to the cockpit side. (Bowers)

A German mechanic spins the propeller on Lieutenant Von Zastrow's Fokker M.14V (E.III), 401/15 at a German Army field near the front. The cowling has been removed, indicating that the crew had probably just completed some sort of maintenance on the engine. (Bowers)

the propeller arc. This threat to the Fokker menace went into operational service with the British Royal Air Service in the Fall of 1915.

A short time later, in February of 1916, the British deHavilland DH-2 went into operational service. Goeffrey deHavilland specifically designed this aircraft to combat the Fokker E

A pair of Fokker E.IIIs are readied for a patrol over the front at a front-line German Army field. The two aircraft have slightly different methods of displaying the fuselage national insignia. (Bowers)

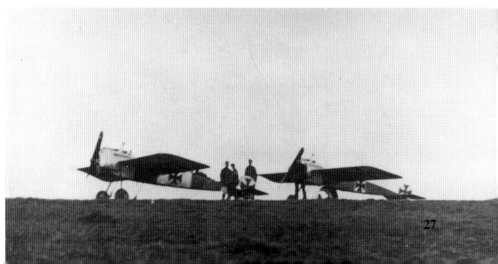

An unknown German Army pilot in the cockpit of a Fokker M.14V (E.III). It is believed that this aircraft is E.III 20/15, an early production aircraft. (Bowers)

The Allied Answers To The Fokker Menace

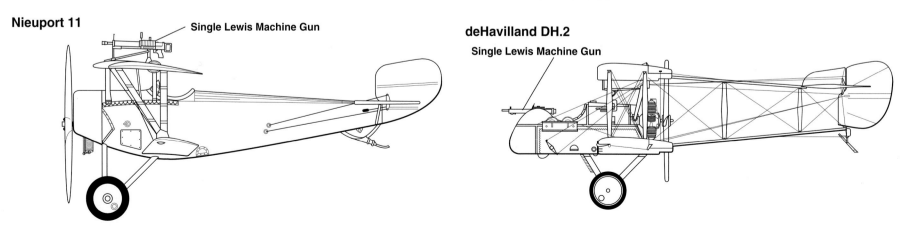

Nieuport 11 — Single Lewis Machine Gun

deHavilland DH.2 — Single Lewis Machine Gun

Machine Gun Sychronization System

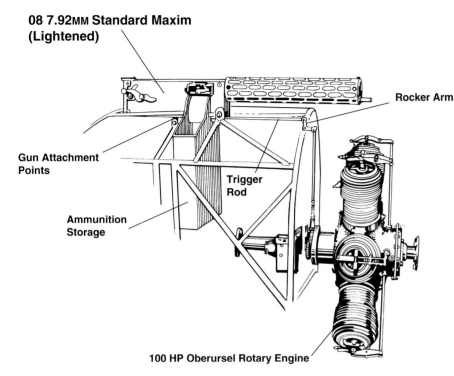

series. The DH-2 was a "pusher" aircraft driven by an 80 hp Gnome rotary. The cockpit, at the very front of the engine pod, mounted a single fixed Lewis machine gun and the gun was aimed by maneuvering the entire aircraft. Extra drums of ammunition were stored outside and around the cockpit area. Fast and very maneuverable, it too helped counter the Fokker threat. By July of 1916, the Allies had re-established air superiority over the Western Front.

For anyone with an interest in the Fokker monoplanes, the Upavon Report Number 48 provides a major source of information on the Fokker E.III. A Fokker E.III, whose numerical designation was not listed in the report, but often referred to as the 210/16 came down behind the lines in April of 1916. The aircraft found its way to England, and during the last three days of May 1916, was tested at the British Upavon facility. A summary of this report follows:

UPAVON Report No. 48

Committee: CCFS, CIA, O.C. 4th Wing, Secretary
Manufacturer's date: 28 March 1916
Engine: 9-cylinder Gnome type, not Monosoupape
Engine serial number: V-1081

Propeller diameter: 8 feet 31/2 inches (2.53 meters)
Instruments: revolution counter, pulsometer gauge, pressure gauge, compass (on wing) aneroid barometer, airspeed indicator (installed by British).
Span: wings, 32 feet 11 inches (9 .99 meters), elevator, 9 feet 6 1/2 inches (2.91 meters)
Overall length: 24 feet (7.32 metes) Height: 8 feet 1/2 inch (2.45 meters)
capacity: main fuel tank: 25.8 gallons (98 litres)
         reserve fuel tank: 5.81 gallons (22 litres)
         oil tank: 6.6 gallons (25 litres)

Particulars (committee's words):

1. Balanced elevator and rudder.
2. Warp control of wings.
3. All wing connections fitted with quick, detachable joints for rapid dismantling.
4. A clamp fitted on the control stick by which it can be locked in any position of fore and aft control.
5. Track of undercarriage wheels 6 feet 8 1/2 inches (2.5m)
6. All steel fuselage-brazed joints - no sockets.
7. Double flap doors in floor between the pilot's knees and shutters on either side under wings, all of which can be worked by the pilot for getting a better view.

Consumption Trials (at Upavon, 5/30/16)

Weight of pilot: 180 pounds
Length of flying time: 10:50 am to 12:07 pm -- 1 hour 17 minutes
Consumption: fuel: 12 1/4 gallons 9.5 gallons per hour
            oil: 3 gallons 2.3 gallon per hour
Engine revs: 1,140 to 1,180 (on ground)
Average air speed: 71-72 miles per hour
Time to 8,000 feet altitude: 17 minutes

Three Fokker M.14Vs (E.IIIs) lined up for inspection on a German Army field. It is believed that Fokker was in the group of men standing between the two aircraft on the right. (Bowers)

This aircraft was being used at the Fokker factory to prepare photographic documentation. The striped pole under the aircraft was used as a measurment scale. Although the aircraft has the serial 419/15 on the fuselage, the fin has a small 401 (probably the factory work number) in Black under the national insignia. (Bowers)

Speed Test (Upavon 5/30/16)

Average maximum speed for three different upcourse and return - 86.4 miles per hour at 1,210 rpm. Minimum speed was approximately 50 miles per hour.

Climb To Altitude Test (Central Flying School)

ALTITUDE(ft)	TEMPERATURE	RPM	MPH
0	15	1,210	86.5
1,000	10	1,205	87.5
3,000	6	1,180	86.0
5,000	5	1,140	84.0
7,000	4	1,140	82.5
9,000	3	1,130	80.0
11,000	0	1,120	77.5

Rate Of Climb Test (Upavon 5/29/16)

ALTITUDE (ft)	MINUTES	SECONDS	AVG FT. PER MIN	REVS	AIR SPEED
1,000	1	0	1,000	1,170	60
2,000	2	45	570		
3,000	4	25	600	1,160	57
4,000	6	20	520		
5,000	8	30	460	1,156	55

This uncovered Fokker E.III airframe reveals the placement of the pilot's seat, control stick, gas tank, engine and ammunition storage bins. (Grosz)

6,000	11	10	370		
7,000	14	0	350	1,100	54
8,000	17	30	290		
9,000	23	0	200	1,080	53
10,000	28	0	180		
11,000	35	0	140	1,060	53
12,000	46	0	80		
12,200	51 (ceiling)				

Further Remarks:

This machine can be dived very steeply and showing an airspeed of 115 mph comes out of the dive with complete ease.

General Remarks:

1. Unstable latterally, longitudinaly and directionally.
2. Amount of vibration in air is very little indeed.
3. All controls are convenient to pilot.
4. Tiring to fly in all but still air.
5. Length of run: to unstick is 75 yards; to pull up, engine stopped, 80 yards
6. Ease of landing: easy
7. Time to ready engine for starting is 3 minutes
8. Other remarks: machine persistently flies right-wing down. impossible to cure without setting springs to control stick. (Author's note: this was probably because of the natural tendency of the aircraft to rotate about the roll axis in the same direction that the engine rotated - caused by the slight friction concomitant with the engine support bearings and air drag within the cowl.)

A damaged Fokker M.14V (E.III) awaits repair at the Fokker factory. The rudder is bent over and the wing tips are creased. The aircraft appears to have a dark painted fuselage. (Fokker)

Suggestion For Improvement

1. Needs mechanical air pump.
2. Wing covering removed between fuselage and first wing rib for a better downward view.
3. Add tail fin to improve directional stability.
4. Engine cowling better streamlined.
5. Airscoops for forced air induction into the engine. Air intake in cockpit only.

Chief Features That Recommend This Type Of Machine

1. Simplicity of fuselage and its construction. There are no lugs or sockets involved.
2. Ease of dismantling and lugs for wings to attach to fuselage for transport.
3. Wings are protected - there are knobs to eliminate leading-edge damage. Entire edge of wing has bound, hardwood covering for protection.
4. Control stick is well designed with engine cut-out and gun trigger where pilot grips the two hand-grip stick. Also locking lever.
5. Compass on right wing is easily seen but probably affected by metal in fuselage.
6. Instruments are all suitable and well placed.
7. Pilot's seat is adjustable fore and aft and up and down.
8. Wheels are well separated.
9. Tail skid is fixed but should be shearable- all scouts have this problem and it complicates taxiing and landing in confined spaces.
10. Observation trap doors are a good idea and well fitted. The British should do as well in their work.
11. Gun sights are difficult for the pilot to use because of the slipstream from the propeller.
12. Fabric is very poor and probably heavy. Germans unable to grow flax and manufacture quality linens as in British Isles because of bad climactic conditions.
13. Miscellaneous: all wire and turnbuckle work finished off very nicely showing greater attention to detail than British fabricators.

A Fokker M.14V (E.III) of Kampf Einsitzer Kommando Vaus parked on a German Army air field near the front. The aircraft in the background is a Pfalz E. II . The Fokker carried the serial 635/15. (Bowers)

This Fokker M.14V (E.III) was flown by Max Immelmann while assigned to KEK III at Douai during 1916. The aircraft in the background is believed to be his Fokker M.15 (E.IV). (Bowers)

Lieutenant Student poses in front of his Fokker M.14V (E.III) of Kampf Einsitzer Kommando III, Armee Vouziers during 1916. (Bowers)

The pilot and ground crew of a Fokker M.14V (E.III) 401/15 pose with their aircraft on a German Army front line base. The aircraft was assigned to Lieutenant Von Zastrow. (Bowers)

Lieutenant Kurt Wintgens takes off in a Fokker M.14V (E.III) from the Fokker factory air field. The aircraft is unusal in that it does not have a serial number painted on the fuselage. (Bowers)

There is only one Fokker E.III known to exist at this time. It hangs from the ceiling in the Aeronautics section of the National Museum of Science and Industry, South Kensington, England. Most historians believe that this E.III is the one that was tested at Upavon and had the numerical designation 210/16. It appears that this E.III has fallen into disrepair since 1916. The gun sights are gone and the engine is not the original Oberursel that the aircraft mounted. All of the fabric has been stripped off. The machine gun synchronizing gear has apparently been removed, or at least the activating cam and push rod behind the engine. It now has a Le Rhone-type propeller for a 110-hp rotary. The carburetor/air intake unit attached to the engine appears, by the Author's observation, to be from a LeRhone engine, rather than the perforated air intake normally installed. It also sports "Dunlop Universal" tires! No apparent structural changes have been made, however, although a bend in the control tube to which the stick attaches is said to have been done when the aircraft was being lifted at the museum.

A Fokker workman attaches fabric to the fuselage of a M.14V (E.III). This is a late production aircraft fitted with a 08 Standard machine gun with a barrel recoil booster used to improve its rate of fire. (Grosz)

(Right) The Fokker M.14V (E.III), like all Eindecker models, used metal tube construction instead of wood, which was common on First World War fighters. Visible are the pilot's seat, control stick, gas tank, ammunition bins, engine, and engine firewall. (Grosz)

The uncovered starboard wing of a Fokker M.14V (E.III) reveals the construction techniques used by Fokker. (Grosz)

The pilot (left) and ground crew pose with their Fokker M.14V on a forward airfield in France during 1916. (USAF Museum)

(Left) This is believed to be the last production Fokker M.14V (E.III). The final lot of E. IIIs produced at the Fokker factory had been initially ordered by the German military in February of 1916. The serial number is belived to have been 249/16. (Fokker)

Lieutenant Buddecke poses with his Fokker M.14V (E.III) in Turkey. The Black square insignia became the national marking of Turkey (later the Black was changed to Red). Initially the insignia was formed by squaring off standard German crosses. (Bowers)

French troops inspect a Fokker M.14V (E.III) downed behind Allied lines. The wheel covers appear to be half and half Black and White, similar to the markings of *Jasta 9*. (USAF Museum)

Workmen apply the national insignia to the wing of a Fokker M.14V (E.III) at the Fokker factory. (Author)

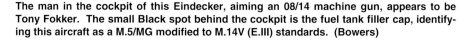

The man in the cockpit of this Eindecker, aiming an 08/14 machine gun, appears to be Tony Fokker. The small Black spot behind the cockpit is the fuel tank filler cap, identifying this aircraft as a M.5/MG modified to M.14V (E.III) standards. (Bowers)

The air park at the Fokker factory in Schwerin. A number of Fokker M.14Vs (E.IIIs) are parked just inside the fence and the aircraft on the grass include an E.III (left) covered in a clear cover intended to make the aircraft less visible. This technique was tested, but found to be of little value and was abandoned. (Fokker)

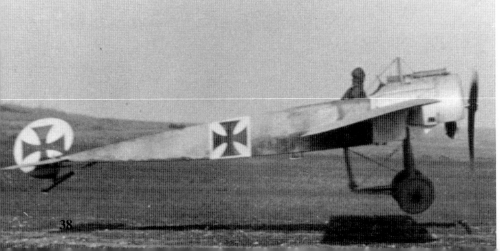

(Left) A rather weathered M.14V (E.III) taking off from a German airfield. The aircraft carries no serial on the fuselage, indicating that it has probably recently undergone an overhaul and has not been fully repainted. ((Bowers)

Fokker M. 15 (Eindecker IV)

There appears to be no overriding reasons, such as those that caused the jump from the E.II to the E.III, to cause the Fokker Company to produce the M.15 (E.IV). The prototype for this production aircraft came out of the factory in November of 1915. The company was filling orders for the E.III as quickly as they developed, with the only slowdown in production being caused by delays in the delivery of 100 hp Oberursel engines. Parchau, a pilot with the German military, flew the E.IV in this same month. He wrote a favorable report on the flight characteristics, and the military ordered the E.IV into limited production.

Kruetzer and staff designed the Fokker E.IV with several major differences between it and the earlier Fokker E.III. Fokker installed a rotary engine with two banks of cylinders, essentially two seven-cylinder Oberursel rotary engines bolted together, giving the engine fourteen cylinders and a rating of 160 hp. To compensate for the additional weight in the front of the aircraft, Fokker lengthened the fuselage to approximately twenty-four feet, eight inches (7.5 meters) to maintain center of gravity. The height increased to about 9.1 mostly from the reinforcing of the main landing gear struts to support the added engine weight. Support bearings were added to the engine mounts so that they circled around the engine to the front, providing front and back support. Engine cooling was improved and the upper wing support pylon was strengthened.

The E.IV mounted two 7.92MM Spandau standard '08 lightened machine guns on the upper cowl, with an improved ammunition feed system. A rudimentary top decking was added that partially enclosed the machine guns and extended past the cockpit, tapering down to the box-girder fuselage structure. The wing and tail spans remained the same as the E.III, along with the wing chord, which remained constant throughout the E series.

It appears that Fokker Company planned for this machine to be a bigger, stouter, more heavily-armed and faster aircraft. Unfortunately for the Fokker Company, the E.IV turned out to be bigger, stouter and more heavily-armed, but also less reliable, slower and much less maneuverable than its E.III predecessor. Most of the problem lay in the engine. The 160 hp Oberursel lost power at altitude and after some time in service, performance tended to deteriorate further, giving the aircraft a poor rate of climb and heavy controls.

Only the best pilots could fly the E.IV because of its clumsiness. The air service issued Max Immelmann a customized E.IV (serial 189/16) with three synchronized Spandau machine guns. He took this aircraft into action for the first time in February of 1916. Immelmann did score three victories with the three-gun E.IV, but reportedly did not like the aircraft. He then went back to the standard two-gun E.IV version. He shot away his propeller and crashed to his death, supposedly in an E.II or E.III, on 18 June 1916.

Oswald Boelcke helped seal the fate of the E.IV with a report detailed in *Air Aces of the 1914-1918 War*. Boelcke stated that the E.IV was less maneuverable and its rate of climb was reduced when compared to earlier Eindecker versions. He stated that he had lost several Nieuports because of the poor rate of climb of the E.IV. A summary of his report listed the following items:

1. The speed of the 160 hp Fokker is sufficient in level flight. In climbing it loses speed.
2. Ability to climb over 9,842 feet (3,000 meters) is insufficient.
3. Maneuverability is less than that of the 80 or 100 hp E types. Quick turns are impossible without stopping the engine. In stopping the engine, one loses height and in combat this would be dangerous.
4. Performance of the engine was good at first, but after use in service, engines lost more than 100 rpm.
5. Machine guns work well under good maintenance, Workmanship and materials in new aircraft cause various troubles.
6. The installation of 15 degree upward tilted machine guns is useless. It would be better to use the old installation with the guns installed in the direction of flight.

He ended his report by suggesting that biplanes might be more suitable! By the Summer of

Fuselage Development

M.14V (E.III)

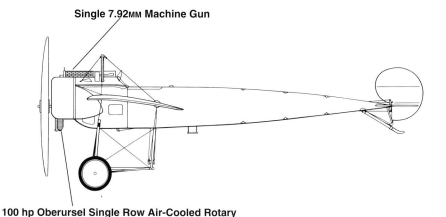

100 hp Oberursel Single Row Air-Cooled Rotary

M.15 (E.IV)

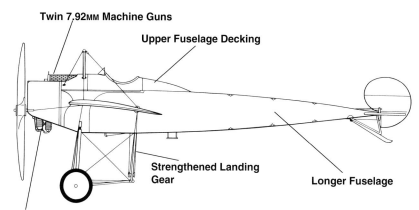

160 hp Oberursel Double Row Air-Cooled Rotary

Oberlieutnant Oslwad Boelcke in flight over the front in his Fokker M.15 (E.IV). It appears that his Fokker had a dark painted cowling, although this can not be confirmed. Boelcke wrote a technical report on the E.IV detailing its shortcomings and causing an order for some twenty aircraft to be cancelled. (Bowers)

1916, it had become obvious that the days of the Fokker Eindecker series were over and that its replacement with better performing fighters was long overdue. Fokker Eindeckers still held the edge on the Eastern front and remained unchallenged until late 1916. After Immelmann's death on 18 June, all Fokker monoplane fighters were either sent to training units or dispatched to the Eastern Front.

Historian P. M. Grosz lists a total of forty-nine Fokker E.IVs being produced until March of 1916. A major order for twenty E.IVs in April of 1916 was cancelled, probably as a result of Boelcke's negative report.

(Left) Lieutenant Hans Müller in the cockpit of his Fokker M.15 (E.IV).(serial 161/16) on 25 April 1916. Lieutenant Müller was assigned to *KEST Boem-Hangelar*. The aircraft has a row of cooling air inlets around the cowling to improve cooling for the double row 160 hp rotary engine. (Bowers)

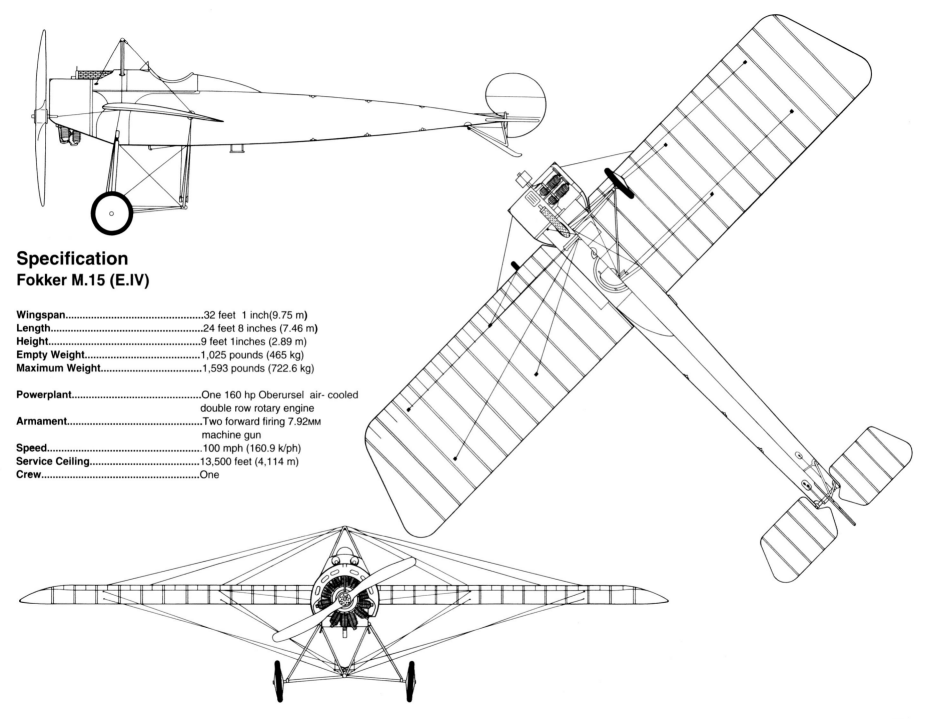

Specification
Fokker M.15 (E.IV)

Wingspan	32 feet 1 inch (9.75 m)
Length	24 feet 8 inches (7.46 m)
Height	9 feet 1 inches (2.89 m)
Empty Weight	1,025 pounds (465 kg)
Maximum Weight	1,593 pounds (722.6 kg)
Powerplant	One 160 hp Oberursel air-cooled double row rotary engine
Armament	Two forward firing 7.92mm machine gun
Speed	100 mph (160.9 k/ph)
Service Ceiling	13,500 feet (4,114 m)
Crew	One

A Fokker M.15 (E.!V) 189/16 on the grass at the Fokker factory airfield. The E.IV was larger, more robust and more heavily armed than any earlier E type. It also was slower and far less maneuverable than the E.II and E.III. (Fokker)

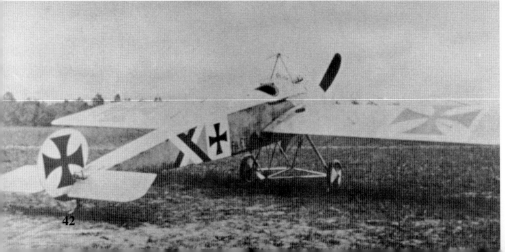

This Fokker M.15 (E.IV) was painted with crossed chevrons (believed to be Black and White) on the fuselage and quadrant-painted wheel covers. It is believed that this aircraft was flown by Max Immelmann. (Bowers)

This Fokker M.15 (E.IV) was set up at the factory to shoot a number of documentation photos of the three machine gun installation. Although Boelcke tested such a gun installation, he reportedly disliked it. (Fokker)

Tony Fokker in the cockpit of a three gun armed Fokker M.15 (E.IV). Fokker flew a number of test flights with this aircraft and on one flight the synchronizer failed to function properly and Fokker put some sixteen rounds into his propeller, nearly shooting it away. Somewhat shaken, Fokker was able to make a safe landing. (Bowers)

A group of German pilots pose with one of their unit's Fokker M.15s (E.IVs) on a front-line airfield. The Eindecker was a strong aircraft easily able to support the weight of the two men setting on the wing tips. (Bowers)

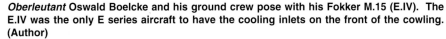

Oberleutant Oswald Boelcke and his ground crew pose with his Fokker M.15 (E.IV). The E.IV was the only E series aircraft to have the cooling inlets on the front of the cowling. (Author)

This Fokker M.15 (E.IV) had a three gun installation angled to fire upward. The bar under the guns was the firing control for the three weapons. The added weight of the three gun installation made it less popular than the standard two gun installation. (Grosz)

A line-up of twenty Fokker M.15 (E.IV) fuselages outside the Fokker factory at Schwerin awaiting wings. (Fokker)

A factory documentation view of a three gun Fokker M.15 (E.IV). The bar visible inside the cockpit was the firing control for the three guns. (Grosz)

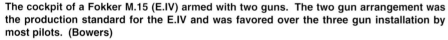

The cockpit of a Fokker M.15 (E.IV) armed with two guns. The two gun arrangement was the production standard for the E.IV and was favored over the three gun installation by most pilots. (Bowers)

This three-gun Fokker M.15 (E.IV) is unusual in that it does not have the cooling intakes around the front of the engine cowling. (Grosz)

Fokker M.15s (E.IVs) under construction in the Fokker factory at Schwerin. There are at least fifteen aircraft in various stages of assembly. (Bowers)

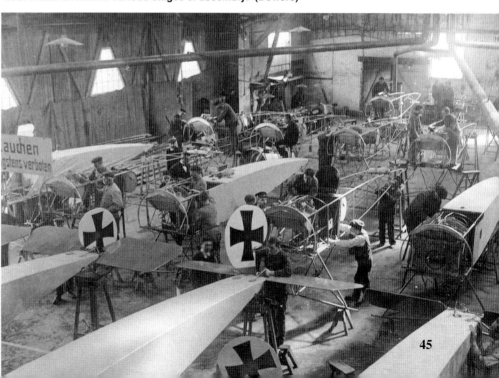

Lieutenant Student poses with his Fokker M.15 (E.IV) during June of 1916. He was assigned to *Jasta 9* and this unit used half Black, half White wheel covers as a unit marking. (Bowers)

(Left) This Fokker M.15 (E.IV) was assigned to Flieger Abteilung 19 during 1916. The pilot is unidentified. (Bowers)

Oberlieutenant Oswald Boelcke in the cockpit of his Fokker M.15 (E.IV) prepares to start the engine. The ground crewman spins the propeller after the engine is primed and the ignition engaged. Boelcke's *Jasta* (*Jasta 2*) used tri-colored wheel covers as a unit marking. (Bowers)

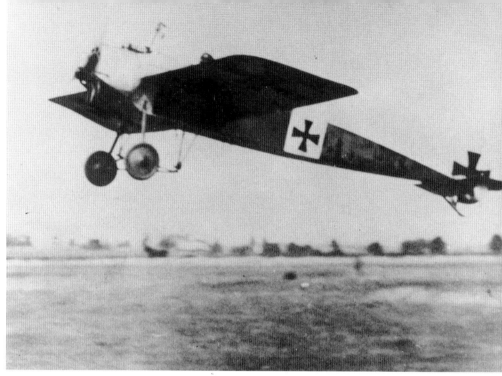

A Fokker M.15 (E.IV) takes off from a German Army field. The pilot and unit are unknown. (Bowers)

A line-up of Fokker M.15s (E.IVs) of the German 5th Army. These aircraft were based at Vouziers, France during 1916. There are both Fokker E.IVs and E.IIIs in this group. (Bowers)

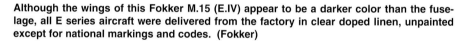

Although the wings of this Fokker M.15 (E.IV) appear to be a darker color than the fuselage, all E series aircraft were delivered from the factory in clear doped linen, unpainted except for national markings and codes. (Fokker)

German Navy Eindeckers

Very little has been written about the use of the Fokker E series by the German Navy, the Marine-Flieger-Abteilung. The Navy used a number of Fokker E types, including E.Is, E.IIIs and E.IVs as land based fighters. Seven pilots were sent for training at the Kampfeinsitzer-Schule at Mannheim. They and their Fokkers were used operationally on the Eastern Front, and of the seven naval pilots trained in fighters, only two survived the war.

This was the first Fokker E series aircraft purchased by the German navy. This aircraft was assigned to the Flanders area. (Bowers)

This variant of the M.5 was designated the M.6. The aircraft crashed shortly after being built and was never put into production. (Grosz)

Other Fokker Monoplanes

The term Eindecker means one wing, or monoplane. Besides the fighter E series, Fokker produced a number of other monoplanes for a variety of roles. Some reached production, while others were just experimental aircraft.

The M.6 used a parasol wing arrangement. Basically a modified M.5L, the aircraft proved to be unstable. (Bowers)

The shoulder winged Fokker M.8 (A.I) went into production during September of 1914 as a artillery spotter, trainer and reconnaissance aircraft. This M.8 was serialed A.96/14. A total of thirty-five aircraft were built. (Bowers)

This Fokker M.8 (A.I) artillery spotter, serial A.210/14 was powered by a captured French 80 hp Le Rhone radial engine. (Bowers)

Fokker M.8s (A.Is) under construction in the Fokker factory at Schwerin. The M.8 was a artillery-spotter aircraft with side-by-side seating for its two man crew. The aircraft was some forty percent wider than the Fokker E series fighters. The aircraft were used operationally by *Feldfliegerabteilungen* 40 and 41 until late 1915. (Fokker).